Edward Nicoll Dickerson

Joseph Henry and the Magnetic Telegraph

An Address Delivered at Princeton College, June 16, 1885

Edward Nicoll Dickerson

Joseph Henry and the Magnetic Telegraph
An Address Delivered at Princeton College, June 16, 1885

ISBN/EAN: 9783337165161

Printed in Europe, USA, Canada, Australia, Japan

Cover: Foto ©berggeist007 / pixelio.de

More available books at **www.hansebooks.com**

JOSEPH HENRY

AND THE

MAGNETIC TELEGRAPH.

AN ADDRESS

DELIVERED AT PRINCETON COLLEGE, JUNE 16, 1885,

BY

EDWARD N. DICKERSON, LL.D.

"SI MONUMENTUM QUÆRIS CIRCUMSPICE."

NEW YORK
CHARLES SCRIBNER'S SONS
1885

COLLEGE OF NEW JERSEY,

PRINCETON, N. J., June 19, 1885.

MY DEAR SIR:

Immediately after the delivery of your paper on the discoveries of Dr. Joseph Henry, I expressed my strong personal desire to see it published.

I find that the College is at one with me in this wish.

We should like to have so comprehensive a paper circulated in our day, and handed down to posterity.

I am yours ever,

JAMES McCOSH.

EDWARD N. DICKERSON, LL.D.

NEW YORK, June 22d, 1885.

MY DEAR SIR :

I take pleasure in furnishing to Princeton College my address presenting the memorial tablet of Professor Henry. I have added to it an appendix of notes supporting the statements contained in it, which I trust will prove satisfactory.

It is a labor of love for me to do anything tending to present our great and beloved friend to his own age, and to posterity, in his true proportions. His achievements seemed to him so easy to perform, that he never looked upon them as exhibiting any great power ; and he therefore involuntarily shrank from that praise to which he was so eminently entitled, and from conspicuously exhibiting his results before the world. He preferred to defer to the judgment of posterity, and to submit his reputation to the ordeal of time, which, like a simple acid, eats away the baser metal, and leaves the pure gold free from its association.

With many thanks to you for the compliment implied in your request,

I am, sir, very truly,

Your obedient servant,

EDW. N. DICKERSON.

President McCosh,

Princeton College.

MEMORIAL ADDRESS

DELIVERED BY

EDWARD N. DICKERSON, LL.D.,

PRESENTING TO PRINCETON COLLEGE A TABLET DESIGNED TO
COMMEMORATE THE CONTRIBUTIONS TO THE
ELECTRIC TELEGRAPH OF

JOSEPH HENRY.

Mr. President and Gentlemen, Trustees of the
College of New Jersey:

The pleasing but sad duty has been assigned to
me of presenting to you this memorial tablet of
the beloved master, who once shed the lustre of his
genius over this ancient seat of learning, and once
attracted to its classic shades, allured by his great
reputation, pilgrims from all lands, to drink from
the living font of knowledge, ever replenished
and refreshed by his ceaseless contributions.

I commit this monument to your tender care.
May it ever remain enshrined in this beautiful
temple. May its presence encourage those, and
the successors of those, to whom he delivered his
torch of science, ablaze with a light which had

penetrated to the farthest ends of the earth, to tend that sacred flame; so that when they shall transmit it to their successors, it shall still be borne high aloft in the upper atmosphere of pure truth, with still increasing lustre — a guiding beacon to the wayfarer, wandering and astray in the gloomy valleys of ignorance—those deep defiles, where the shadows seem ever darkening by contrast with the brightening mountain tops illumined by the rising sun of knowledge.

May it inspire the ingenuous youth, who in the thronging years of the future shall gather about these altars, to search the character and achievements of the great master; that they may be taught by him how to study; how to think; how to work; how to live; and how to die.

May it continue to remind those who annually are attracted here to witness the evidences of the growth of knowledge, as they are exhibited in the commencement seasons, that once this college was honored by the ministrations of Joseph Henry, an American, who, with means created almost wholly by himself, rivalled the achievements of the greatest scientists of the old world, working with the resources of nobly-endowed institutions, and encouraged by the bounty of Kings; and for years was ever a leader in the vigorous attack upon the arcana of nature, made by the champions of science in the early years of this century.

For those of us who enjoyed the happiness of knowing him well, and loving him dearly, no sculptured marble is needed to stir our hearts, or keep fresh in our memories that noble presence, which at once charmed and satisfied our senses. Nor, if the chisel of the artist were guided by the genius that once inspired Phidias, would it be capable of fixing upon dull, cold marble more than one of the almost infinite variety of expressions revealing to the world without the exalted being within.

But to those who have never seen him, or having seen him have never known him, and to those who shall come after us, it will be something to look upon this marble, and inspired by the thoughts he uttered, and the deeds he did, contemplate its calm expression, and imagine what must have been the living man.

In the year 1839, nearly half a century ago, brought here as a student, I first saw Professor Henry. I remember it well—the time, the place, and the surroundings. Boyish imagination had pictured the great discoverer as a venerable man, bowed down with the toil of years, bearing the furrows, with which overtasked nature revenges herself, traced upon his brow : such a person, perhaps, as the artist has presented to us in the familiar picture of Humboldt in his library.

How different the reality ! In the maturity of a
perfect manhood he stood :

> " A combination, and a form indeed,
> Where every God did seem to set his seal,
> To give the world assurance of a man."

His clear and delicate complexion, flushed with
perfect health, bloomed with hues that maidenhood
might envy. Upon his splendid front, neither
time, nor corroding care, nor blear-eyed envy,
had written a wrinkle, or left a cloud ; it was
fair and pure as monumental alabaster. His
erect and noble form, firmly and gracefully
poised, would have afforded to an artist an ideal
model for an Apollo. The joy of conflict and of
triumph beamed from his countenance—a conflict
in which, for years, he had struggled with the
phantoms that guard the hidden treasures of
nature, and had ever been victorious. And above
all, surmounting all, infinite charity and gentleness—like the charity and gentleness of a loving
mother for her erring children.

To him the youthful student bowed down in
profound admiration. To him, and to his memory,
for nearly fifty years, he has clung with ever-in-
creasing love and affection. And now that seven
years have passed away since death severed the
bond strengthened by a life-time of intimacy,
he recurs with fondest memories to the many
happy occasions when it was his good fortune

to spend hours in sweet and instructive converse with this gifted mortal, to whom the whole book of nature was an open volume, out of which he ever read lessons of wisdom, and beauty, and truth.

As Professor Henry appeared in 1839, so he continued till 1847, with but little change in the physical man—only that change, which, like the changes in the early autumn, lightly touched with tints of exquisite beauty the mature growth of spring-time and summer ; and then, with extreme reluctance, he departed from Princeton, called by his country to lay down the arms with which, as a soldier in the ranks, he had been waging his warfare against ignorance, and take command of the intellectual forces to be summoned and organized by him in the same glorious cause.

Born in the dying moments of the eighteenth century, his age was marked by the numbers denoting the years of the nineteenth. Like the century, with whose growth his growth kept pace, he had developed with almost unexampled rapidity; and at the age of thirty-two, when he took his chair here, although "he was but a "youth, and ruddy, and of a fair countenance," and was armed only with a simple sling of his own construction, and pebbles from the brook of nature, he was equal to the trained warriors of maturer growth and superior armor, waging war against the Goliah that guarded the unexplored regions of nature's secrets; and like the great king of Israel, after the brunt of the battle was over, he came to be leader of the hosts, who once had been tending only a "few sheep in the wilderness."

Let us contemplate for a moment the intellectual stature of our departed teacher, considered merely as an investigator of natural laws, and measured by the standard established by the intellectual world.

It is in the order of nature that the intermittent progress of humanity is made under the guidance of gifted men, appearing from time to time, who push forward the outposts of truth, whether in morals or physics, calling upon their fellow-men to hasten and occupy the newly-conquered fields. The names of such men are few, and are written upon the rolls of fame. Their glory belongs to no nation, but to all mankind. Sometimes simultaneously and in different parts of the world two such appear, who seem to have been cast in similar moulds, lest perchance one might die or fail, and progress stand still. Such men were Henry and Faraday, whose intellects were moulded with the same capacities, and who worked out their tasks in the same spirit. If either one had died before his work was done, the other was capable of doing it; and, in fact, both in many cases struck out the truth, each unconscious that the twin thought had been born in the brain of the other.

To those devoted friends and admirers of Faraday, who delight in singing his well-earned praises, and who best comprehend his achievements, it seems that his discovery that electricity might be produced from magnetism was his grandest result. Upon it depends many of the most important applications of electricity to the uses of man; and in the near future many more are coming. Tyndall, the successor of Faraday, does not restrain his enthusiasm when he contemplates this achievement. "I cannot help think- "ing," says he, "while I dwell upon them, that " this discovery of magneto-electricity is the " greatest experimental result ever obtained by " an investigator. It is the Mont Blanc of Far- " aday's own achievements. He always worked " at great elevations, but higher than this he never " subsequently attained."*

Let us accept the standard, and apply it to Henry; let the achievement measure the power of the man.

In November, 1831, Faraday read before the Royal Society his memorable paper "On the Evolution of Electricity from Magnetism," illustrated by drawings of the apparatus, in which Figure 1 is the *compound* "spool," dis-

---

* See Appendix, Note A.

covered by Henry in 1828, and which Faraday used in making his discovery.* No publication referring to this paper had reached this country till April, 1832, when a vague reference, made to it in the "*Annals of Philosophy,*" was seen by Henry, which led to his publication of July, in "*Silliman's Journal,*" where he gave a full account of this great discovery, made by himself before he heard of Faraday's work, which, when compared with Faraday's paper of November, exhibits Faraday's experiment for solving the problem. When he wrote his paper, Henry, mislead by the imperfect statement in the "*Annals of Philosophy,*" supposed that his experiment had differed from Faraday's, but was undeceived when the full publication reached him. In 1831, a teacher in the Albany Academy was very remote from London, and the Royal Society.

In that same year, and in the same few weeks, Faraday first, and Henry after him, independently made the discovery of magneto-electricity; "the "greatest experimental result ever obtained by "an investigator," in the opinion of Tyndall.†

In the same field, and during the same years, were the other great scientists of the world, studying the same subject : Ampère, Arago, Oersted, Davy, and a host of others; but these two did it, and

* See Appendix, Note B.
† See Appendix, Note C.

not the others ; and Henry did it by devices of his own invention, unaided by anything which Faraday had discovered or produced, while Faraday used Henry's electro-magnet in performing his most important experiment.

The towering heights which were scaled by the daring spirit of Faraday from the East, were at the same time surmounted from the West by our own countryman. Both were climbing from opposite sides at the same time, and neither was conscious of the other's efforts till both stood, face to face, upon the summit. Had Henry been furnished with the corps of trained mountain guides, and Alpine-stocks, such as attended Faraday in his ascent, perhaps *his* foot would have first trodden the peak, and Prof. Tyndall's song of triumph have been addressed to him.

But when we compare Henry with Faraday, who is the acknowledged unit of comparison, the accidental conditions under which both existed and worked must be known, or justice cannot be done. Electrical science was the field to which both spontaneously directed their studies. Its mysteries at once excited curiosity, and baffled research. Its most obvious phenomena had only for a short time been recognized, and everything was to be learned. What they did in that science, not only constitutes the greater part of their claims to reputation as investigators, but is almost the whole of our present knowledge of magneto-electricity.

But how superior in every respect, except in God-given intellect, was the equipment of Faraday. He was eight years older than his rival. In the year 1813 he was appointed "assistant" in the laboratory of the Royal Institution, under Sir Humphrey Davy, then one of the foremost scientists of the world, who, attracted by Faraday's genius, was directing his studies and forming his mind. At that time Henry was but THIRTEEN YEARS OLD.

In the next twelve years Faraday was at work, with all the resources of the Royal Institution, under the instruction of the great Davy, in acquiring the knowledge with which he was armed when he began his original investigations; while Henry, during that same period, was struggling unaided for such education as might be obtained from the scanty resources of a country town; and with that proud independence, ever so marked a feature of his character, was supporting himself by teaching to others a part of that which he was learning himself.

In 1825, Faraday had so improved his great opportunities, that at the age of thirty-four he was appointed "director of the laboratory" of the Royal Institution, where everything that science could suggest, and money procure, was at his command in aid of research. Henry was then a private tutor in a distinguished family at Albany ; studying mathematics in hours when his duties to his pupils had ceased, and when other young men might have thought they had earned the right to relaxation and enjoyment.

In 1824, before Henry ever had in his hands any instruments for research in electricity, Faraday, thus trained and equipped, began his attack upon the problem ·of magneto-electricity and failed ; and in 1830 it was not yet solved.

The discoveries of deductive science need no apparatus. They are made and matured in the brain ; and to record them is the only physical incident to their existence or development. Plato would have looked with disgust and contempt upon a laboratory ; and would have scorned the suggestion that time, or place, or physical surroundings, could affect the workings of his mind, or influence his deductions. But the new philosophy, which has changed the face of the world, is of no such ethereal nature. It is born in observation of physical things; it is nurtured upon experiments that cost money, and time, and labor; its maturity is in perfected arts, and in things to be seen, and handled, and enjoyed by the senses; its end is to subordinate the blind forces of nature to the uses of man—to mitigate the ills, and multiply the joys of life. They who are the servants of this philosophy, must be provided with materials with which to reproduce, in miniature, the conditions that exist in nature in grander proportions, or they cannot ask the questions whose answers they are seeking; and, other things being equal, he who is well provided with all these needful things, has an immense advantage over another who lacks them.

For thirteen years Faraday had been pursuing his investigations amply supplied, and was in the full career of successful experiment when, in 1826, his great rival first looked upon the course over which he was to run ; and even then Henry had to depend upon the meagre facilities of the Albany Academy, and the voluntary assistance of an appreciative physician, Dr. Philip Ten Eyck, of Albany—a name to be held in grateful remembrance by all who feel a pride in the achievements of the great scientist, whose early efforts were assisted, and whose hopes encouraged, by this enlightened friend.

With such a beginning as this, who could expect that the young aspirant for fame should ever overtake his great leader in the friendly contest ? And when he did overtake him, and in some important investigations surpass him, who shall deny that Henry, as a physical investigator, was the equal of him above whom it is conceded no other man has risen in this century!

In still further pursuing his researches into the subtle phenomena of electricity, Henry made, here in Princeton, another capital discovery, this time in advance of Faraday, which forms an important element in the science of electricity. It is to be found detailed in any school book, under the name of " Henry's Coils." His wonderfully elaborate investigations will be remembered by the students of that day; as it was conducted in part in the open air. Wires stretched across the campus, in front and in rear of Nassau Hall, were the means by which the questioner was cross-examining nature, and wresting from her reluctant grasp her hidden secrets. At that time telegraph wires did not exist; and those fine lines traced across the sky, excited the liveliest interest in the students, whose fantastic guesses as to their significance were the cause of much pleasantry in the idle hours.

In the course of these investigations it was also the good fortune of our scientist to first discover the very curious phenomenon of "self induction," as it is now called, which plays so important a part in the creation and use of electric currents on wires, sometimes injuriously and sometimes beneficially. Without the knowledge of its laws no duplex or quadruplex telegraph could be practically operated; with that knowledge it can be neutralized when it is injurious, and made available when useful. The brilliant spark which follows the pulling of the pendant, attached to an electric lighter for inflaming a gas jet, now in common use, is one of the valuable practical applications of this principle so discovered.

In contemplating the discoveries of the scientist there are two aspects in which they present themselves. In one view we consider merely the difficulty of the achievement; in the other, the value of the result to mankind. The first view is obvious when the thing is done; the other is to be reserved for a future day, when all the consequences have followed the original cause. The first view is that which measures the power of the man— just as the lifting of a huge weight by some Hercules exhibits his strength, even though the thing done may be, or may seem to be, useless. The capital discoveries I have named, made by Henry and Faraday, exhibited the giant's strength when they were made, and measured the men who made them. They were found at great depths below the surface, where mental vision can only penetrate by the aid of lenses, constructed *in advance, in accordance with the very laws for whose discovery they are needed*—creations of the scientific imagination, and called scientific hypotheses. In such creations Professor Henry was excelled by no man.

Time will not permit even a hasty review of all

the scientific labor done by Professor Henry at Princeton, during those years when his chief duties were instruction, and when he had only a portion of his time in which to work for mankind and for reputation ; and I must be content with a passing glance at a part of it.

Among those wires which were strung across the campus in 1835, was one used for a magnetic telegraph between the professor's home and his laboratory in the Philosophic Hall; and that telegraph line was the first in the world in which the galvanic circuit was completed through the earth—one end of the single wire-circuit terminating in the well at the house, and the other in the earth at the Hall. Steinheil, in Munich, in 1837, worked his electric telegraph in the same way by a single line wire, using the earth as part of the circuit, over much longer distances; but it was first done in this campus.

Nearly a century earlier our great countryman, Franklin, had drawn from a surcharged thundercloud, upon the string of a kite, in a pouring shower of rain, the lightning of heaven, and had demonstrated its identity with the puny spark of an electrical machine; and with that capital experiment his fame is more closely associated than with any other of the great truths he discovered. In these grounds that experiment was amplified, and still further results obtained, by the man for whom the mantle of Franklin had been waiting all those years, and who was the only American whose stature would not have been dwarfed by assuming it.

From the clear, blue sky, with two kites, one above and assisting the other, held by a delicate wire wound on an insulated reel, Professor Henry drew down streams of brilliant sparks, intensified by the self-induction of the wire itself; thus proving the electrical relations of the earth and its envelope. So, a child's plaything in the hands of a master, reveals the hidden mysteries of the universe.

Away beyond the distant horizon we see at times a quivering illumination of the sky, but hear no thunder. How shall that phenomenon be questioned ? Fifty years ago, Henry converted the metallic roof of his house into a great inductive plate, by soldering to it a copper wire, and leading that through an electro-magnetic coil to the ground; and with that he held converse with the distant lightning, so far away that its voice could not be heard. If the gods of mythology, who hurl their thunderbolts, have a system in their signals, this apparatus would enable us to read their thoughts. Within a few months, a device has been put into operation by which telegraphic communication is kept up between the running cars on railroads and the stations, so that the positions of all the trains may at any time be known, and protection against collisions assured. To do this the metallic roof of the car is used as an inductive plate, just as was the house roof fifty years ago; and a wire passes from it through a signaling coil to the ground by way of the metal wheels and track. Near the roof outside, an electric wire is stretched on poles, through which electric flashes, like lightning, are sent, and they set up by induction in the roof electric currents similar to those passing over the wire, which are read as signals by the observer; and, conversely, signals are sent from the roof to the wire by induction coils in the car. The experimental demonstration in Princeton has not been lost, though buried so long, and to-day it throws another safeguard around our lives.*

The first electro-magnetic engine for generating power was made by Henry, at Albany, in 1831.† His clear mind was not deluded into the belief that such an apparatus could supersede the steam engine as an economical motor, and he warned the world against that delusion. Zinc, as fuel in a battery, is more costly than coal in a furnace. Still, he saw and said that in exceptional cases it might be useful; a result now coming to pass, dependent, however, upon the discovery of magneto electricity

---

* See Appendix, Note D.
† See Appendix, Note E.

by which galvanic batteries are dispensed with, and electricity, made in quantities from some great and economical source of power, is distributed to Henry's machines wherever they may be.

In many volumes, some of which have perished by fire, and some remain, were laid out lines and plans of investigation by Professor Henry, needing only leisure and means for their development, covering fields where other investigators have since reaped rich harvests of fame, but from which he was debarred by the pressure of his other occupations here. In those records are contained the evidence that the great intellect, which did so much with so little, was capable of grasping the whole circle of physical science, and of enriching and adorning any department of it to which his efforts might be directed.

But he was destined for another career. A benevolent Englishman, inspired by the noble ambition to aid in elevating mankind, had bequeathed to the United States a great sum of money to be used for "the increase and diffusion of knowledge among men." It was a splendid gift, and a sacred trust. Who was to be found equal to the task of effecting this grand purpose? The civilized world was interested in that question. Mankind was the beneficiary of the trust; and all men were entitled to be considered in its administration. By the common consent of the wisest and best of Europe and America, Professor Henry of Princeton College was selected, and solicited to assume that onerous duty. What tribute was that to the achievements, the attainments, and the character of the man ! He must be famous, that his selection might at once command the assent of the world ; he must be learned, that he might be able to carry out the purposes of the donor ; and he must be virtuous, that he should not degrade the high office to any base or selfish uses. And thus he was called.

When brought to the parting of the roads, choice was extremely difficult. On the one hand, a life devoted to the most delightful of all pursuits—the searching out the laws of nature, which are the thoughts of God ; a reputation already great and daily growing ; and a happy home, surrounded by congenial and loving friends, and undisturbed by cares for the present or the future. On the other hand, an abandonment of the field of scientific research, where the harvest was abundant and the laborers few ; and a surrender to others of the prizes he saw glittering before him in the race he was running ; and furthermore, a grapple with the problems of organization and finance, and with the discordant elements which the scheme of the Smithsonian Institution would necessarily evoke. He foresaw that he *might* find himself, after some years had passed, like a giant shorn of his

strength; on the one side outrun in the race where
he had ever been in the lead; and on the other so
hampered and crippled as to be unable to accom-
plish the great objects for which alone he was
about to abandon his first love. That high sense
of duty which governed him in every act of his
life decided the question, notwithstanding his firm
conviction that in accepting the trust he left the
happiest days of his life in the past.

Perhaps he might have decided otherwise if
Princeton College had been then as it is now. Perhaps he then might have felt, that with such
ample resources at his command as are now to
be found here, his services to humanity might
be greater as a soldier in the ranks than as a
commander in the field.   But at that time no one
had arisen among the friends of this institution
who, like the Medici of the fifteenth century, was
able at the same time to gather the wealth of the
world by the arts of honorable commerce, and to
appreciate that the gathered wealth of the world
owes its existence and preservation to science, to
art, and to literature ; and that therefore it is due
to education that it should be encouraged by
noble gifts, such as have enlarged the capacity of the College of New Jersey, and re-
flected honor upon the names of those whose generous hearts, guided by wisdom, have led them to
broaden these ancient foundations, and to arm

with improved facilities the workers who are here devoting their lives to the advancement of knowledge. All honor to such men.* Had such assistance come earlier, the career of the great scientist might have been different; but it was not to be, and thenceforth another life opened before him, and another man was unfolded to the world.

Perhaps the highest praise that can be bestowed upon any man, is to say of him that he is *just equal* to all the duties ever imposed upon him, and never above them; that his reserves are not called into action until the emergency requires them. Such men are the great benefactors of mankind. Such a man was the Secretary of the Smithsonian Institution. The principles he laid down for the administration of the noble gift of Smithson required time for their development, and promised no present brilliant results. The foundations were to be laid deep in the earth, where the laborer and his work were scarcely to be seen by the passer-by. No popular applause would greet the achievement for years to come, while popular clamor was ever ready to cry out against the waste of time and money that produced no instant fruits. The man of clear purpose and resolute will stood guard over the work; and with just force enough, and no more, drove off the assailants till the foundations

* See Appendix, Note F.

were all secure, the superstructure begun, and it was strong enough to stand alone.

With the skill that would have adorned a professional diplomatist, he temporized and compromised, when he could no longer contend with success ; with the dash that would have illustrated a general, he attacked when the moment was propitious, and the adversary off his guard. With the earnestness of sincere conviction, and the directness of demonstration with which his scientific training had armed him, he convinced, one by one, those who opposed his views, until at last the Regents of the Institution, and Congress, surrendered their judgments to his, and the field was won.

A great library was the dream of Mr. Choate, the most scholarly and persuasive of advocates; and, as a regent, he possessed and wielded a formidable power. It was hard to persuade him that a library does not "*increase* knowledge among men," and that it is very likely to "*diffuse*" ignorance. To discover and accumulate *new* truths, and to diffuse them over the whole earth, was the Secretary's conception of the donor's intention; to pile up in Washington a miscellaneous collection in print of *old* truths and *old* errors, was the idea of the scholars; and they were so strong that a temporary compromise was necessary. The vigorous growth of the true conception at last overshadowed the false one, and the library no longer saps the life of the Institution. Professor Henry always thought that over every library portal should be written some such warning as— " Cave Canem "—beware of the lies.

It was not till 1852 that the serious attacks upon the Smithsonian came to an end. On the 24th of June, of that year, a United States Agricultural Convention met in the theatre of the Smithsonian building. The plan to plunder the Institution seems to have been carefully considered and matured ; and the officers of the Smithsonian were elected members of the convention. Stephen A. Douglas was at that time at the height of his power. He had risen from the ranks by the arts of the politician, and was the most influential man in the Democratic party of that day. Although not yet forty years old, he had just succeeded in defeating General Cass in a contest for the presidential nomination at Baltimore ; and although he failed by a few votes to secure it, he had thrown it to Franklin Pierce, of New Hampshire, and thus kept it open for himself in 1856, as the Western candidate of the party.

He was styled the " Little Giant "—not in derision, but in admiration; as expressing the combination of a small stature and great intellect. Representing in the convention what was then an almost entirely agricultural constituency, he thought that votes were to be got, and his influence strengthened, if he could bring home to them the spoils of the Smithsonian; and accordingly a resolution was introduced petitioning Congress to appropriate a portion of the Smithsonian money for an agricultural bureau ; and Judge Douglas undertook the congenial task of accomplishing the raid.   The recollections of that battle are among the valued treasures of memory, associated in my mind with Joseph Henry.   In such an assemblage, and with such a cause, Douglas was an adversary to be feared by any man. That he was an accomplished politician was proved by his great success ; and he was there to fix another step in the ladder by which he had climbed so high.   His speech was adroit, as only he could make it.   Its argument was founded upon the proposition that civilized man depends upon agriculture, without which barbarism would sweep over the land; and his conclusion was that the farmer was entitled to whatever assistance could be got out of the money of Smithson, whose benevolence could best be applied in encouraging those who were at the very foundations of civilization. It would be great injustice to Judge Douglas to assume that he supposed the diffusion of papers of turnip seed among farmers was that sort of " in- " crease and diffusion of knowledge among men " designed by Mr. Smithson ; but no doubt it would be an increase and diffusion of the knowledge that he was the friend of the farmer, and that was of more importance to him. The Secretary, surrounded by a few earnest friends, and prepared for the assault, sat in the back seat of the theatre quite unnoticed, kindling with righteous indignation at this nefarious plot to confiscate the funds of which he was the chosen guardian, and to destroy the institution devised by his intellect, reared by his unceasing efforts, and guarded so far by his sleepless vigilance.

When the popular applause following the "Little Giant's" popular speech had subsided, the Secretary arose. In measured and dignified words he presented himself as the guardian of that fund, bound, so far as in him lay, to defend it from spoliation. He first developed the moral aspects of the question, and appealed, over the head of the advocate, to the honesty of the constituents he represented; expressing the most generous confidence that the farmers of this country would never consciously be parties in an attempt to seize that which belonged to mankind in general, or seek by a forcible partition to destroy the unity and efficiency of the fund.

The legal aspect of the question he next discussed like an equity lawyer; and denounced in scornful sentences that attempted breach of trust which was implied in the resolution.

And then, out of the fullness of his knowledge, with abundance of illustration and example, he demonstrated that the discovery of new truths, and their application to the arts, had elevated the farmers from the mere drudges they were in the seventeenth century to their present high state of intelligence and comfort.

The effect was overwhelming ; and the " Little Giant " must have felt that there was another "giant" there to whose title no diminutive prefix could be properly applied.

The meeting adjourned till the next day, and these significant words were written in the Secretary's diary, under date of June 25 : " Judge " Douglas, toward the close, made an apology for " the warmth of his expressions. I did the same. " Judge Rusk followed—*so the whole was am-* " *icably settled.*"

Since that day no further assaults have been made on the Smithsonian Institution ; and it stands a proud monument to the genius, the learning, the labor, and the character of the great Secretary; who was content to sink his personality in the impersonal institution—to be overshadowed by the creature of his own creation, in order that true knowledge might the better be increased and diffused among men.

The conscientious obligation he felt pressing upon him to lose no opportunity for diffusing knowledge and correcting error, imposed a vast amount of unrecognized and unrequited labor. The intellectually halt, and lame, and blind, continually resorted to him for help, either in person or by letter ; and they never were sent away empty. Like the home of the lovely country parson in the Deserted Village,

> " His house was known to all the vagrant train ;
> He chid their wanderings, but reliev'd their pain."

They who have witnessed some of those deeds of charity, will never forget the gentle patience with which he listened to the beggars for knowledge, and the simple way in which he conveyed to their imperfect intelligences the truths they were seeking. Their self-conceit was often offensive ; but he knew it was the product of ignorance, and his effort was to cure the disorder. He was no more repelled by the disagreeable symptoms, than the physician is who must treat a loathsome disease. On one occasion, in my presence, one of these cripples refused to accept the instructions of the great physicist, on a very simple question of dynamics, applicable to a project he had in hand; but instead of dismissing him, the master quietly took down "Hutton" from the book-case, and patiently read that author's confirmation of the law he had been teaching. What an exhibition of true humility ! Perhaps, thought he, " I can give a new direction to this

" man's mind, who may yet do something useful ;
" and what matters it that he scorns me."

No one can form an adequate estimate of the vast-
ness of his mind, of the extent and accuracy of his
learning, and of his power to discern the correla-
tions of knowledge, who has not carefully read
the instructions mapped out by him for the guidance of investigators working under the auspices
of the Smithsonian Institution.   They constitute a
set of charts, which, for years to come, will guide
the explorer safely and surely in future voyages
for the discovery of new truths ; and are a monument, attesting the fidelity with which the great
trust was executed, and vindicating the sagacity of
those eminent men, who, in 1846, saw, what his
innate modesty forbade him to see, that Joseph
Henry was, of all living men, the most fit to ad-
minister a fund whose object was "the increase
"and diffusion of knowledge among men."

Passing by thus hastily the great achievements
illustrating the long and happy life of Henry,
let us examine with more particularity his connec-
tion with the electro-magnetic telegraph; whose
creation has so largely modified the course of modern civilization, and endowed the dull earth with
nerves, like those of the living frame, whereby the
whole body of mankind instantly feels the joys or
sorrows of any of its members.

How to communicate intelligence instantly, over

distances so great that the voice cannot be heard, had been well known to organized societies from remote antiquity. Visible signals, made by moving vanes by day, and lighted torches by night, were known to Greeks and Romans alike; and more recently the alphabet was associated with these movements, so that alphabetical messages were freely communicated.

Even barbarous nations and tribes possessed this art in a high degree of perfection; and the arrival and progress of Cortes in Mexico were communicated by telegraphic signals, corresponding with the sign language of the Aztecs, to the capital of the doomed Montezuma.

When atmospheric electricity came to be artificially generated, it occurred at once to ingenious men that it might be used for telegraphy; and, in 1774, the first electric telegraph ever constructed was established at Geneva by Lesage. He used twenty-four wires, each connected with an electroscope, whose function it is to move when the wire is charged with electricity, and by means of which any of the letters of the alphabet could be transmitted, by simply discharging a prime conductor of an electrical machine into the wire corresponding to that letter. This complicated apparatus was subsequently improved by using only one wire, and causing lettered wheels to revolve synchronously at the two stations, so that the same letter would appear at the same time to both operators. By this apparatus, whose principle of synchronous revolution is the same as that now used in the printing telegraph, the sender would simply close the circuit on his electrical machine when his revolving wheel presented the desired letter, and the pith-ball electroscope, moving at the receiving end at the same instant, would indicate to the receiver that the letter then presenting itself to him on his wheel was the one intended.

A number of other inventors used static electricity for the same purpose during the latter years of the last century, and the earlier ones of this. In England, Ronalds had a line of eight miles on which the wire was suspended from poles, and insulated by silken strings;* and in 1796, Salva, in Spain, worked a line by static electricity twenty-six miles long.†

In the year 1800 Volta produced the voltaic pile, and gave to the world that new manifestation of electricity called galvanism. In that form this subtle agent is far more manageable than in the form of static electricity; and by the use of galvanic batteries a current of low tension, but of enormously greater power, can be maintained with little difficulty; whereas static electricity

---

* See Appendix, Note G.

† See Appendix, Note H.

is like lightning, and readily leaps and escapes from the surfaces on which it is confined. The galvanic current also readily decomposes acidulated water, and many other substances, and this capacity was soon applied to the purposes of telegraphy. Sœmering, in 1807, invented a telegraph on this plan, and continued it for several years in Munich, publishing accounts of it in scientific journals, and exhibiting it to learned societies.* Others followed his lead, until finally it came into commercial use in England in 1846 as a rival to the electro-magnetic telegraph of later invention ; but requiring its aid, as an alarm.†

In 1820, Oersted discovered the capital fact that a galvanic current, passing through a wire placed horizontally above, and parallel to, an ordinary compass needle, will cause that needle to sway on its axis to the east or west, according to the direction of the current through the wire. At once Ampère suggested the application of the new discovery to the old telegraph, whereby galvanism might be substituted for static electricity, and the deflection of a magnetic needle for the divergence of the pith balls of the electroscope. Baron Schilling, a Russian nobleman, inspired by the love of science, accordingly took up this suggestion, and constructed a galvanometer or needle tele-

* See Appendix, Note I.
† See Appendix, Note K.

graph, which in a practical and operative form was exhibited to the Emperor Alexander in 1824, and came to be well known to scientific persons at that time.*

In 1833, Gauss and Weber set up a *single circuit* galvanometer telegraph on this plan at Gottengen, leading the wire over the house-tops, on insulators, as we do now; and by the deflections of the needle to the right and left made up the alphabet, as it had been done before when using other means for moving the vanes.†

Their apparatus, however, is perfectly silent. The needle is suspended by a thread when the noiseless current sways it to and fro with but feeble force; and it is incapable of calling the attention of the operator to receive its message. These were serious difficulties, to be overcome by other principles, and other inventions, which would supersede this one.

Following Oersted, Arago, in France, in 1820, made the next capital discovery. It was but a little thing he saw—simply that a sewing needle, surrounded by a coil of wire, through which a voltaic current passed, had become magnetic; but that little thing has grown to be mighty. This observation was the complete discovery of electro-magnetism, which had been dimly seen in Oersted's gal-

* See Appendix, Note L.
† See Appendix, Note M.

vanometer; and was the germ of the electro-magnet. For *four years* this beautiful discovery was experimented with by all the scientists in Europe before another step was taken; and then William Sturgeon, of England, produced the electro-magnet. It consisted of a large soft iron wire, bent into a horse-shoe form, coated with varnish, and wrapped with a spiral coil of naked copper wire from end to end, through which the voltaic current might be passed. This bent wire became a magnet while the current flowed, but lost its magnetism when the current ceased.

Here then was born into the world an apparatus capable of exerting a stronger power at the will of the operator, by merely opening and closing the voltaic circuit; and it was then thought that the difficulties in the way of the telegraph were conquered. The experiment was soon tried with Sturgeon's magnet by Barlow, an eminent scientist; and in January, 1825, he published his results in the *"Edinburgh Philosophical Journal,"* in these words:

" The *details* of this contrivance " (a telegraph) " are *so obvious,* and the *principle on which it is* " *founded so well understood,* that there was only " one question which could render the result " doubtful; and this was, is there any *diminu-* " *tion of effect by lengthening* the conducting

" wire?" If not, he proceeds to say: "Then no
" question could be entertained of the practicabil-
" ity and utility of the suggestion above adverted
" to.   I was, therefore, induced to make the
" trial; but I found such a sensible *diminution*
" *with only two hundred feet of wire,* as at once
" to convince me of the *impracticability of the*
" *scheme.*"

Barlow's experiment was repeated by other scientists in that and following years with a like result;
until it came to be accepted in the scientific world
that the telegraph could not be worked with the
newly-discovered electro-magnetism.   So strongly
was this fixed in the opinion of the day, that as
late as 1837—*thirteen years after the invention of
the electro-magnet by Sturgeon*—so eminent a scientist and discoverer as Wheatstone, pronounced
the electro-magnetic telegraph impossible, on an
occasion when the very question was submitted to
him for decision by Cooke, at the suggestion of
Faraday himself.   This fact is so important, and
so conclusive on the question now under examina-
tion, that I read Wheatstone's own account of it,
submitted by himself to arbitrators who were to
decide a controversy between himself and Cooke
as to their respective merits as inventors of
one form of the electro-magnetic telegraph.   He
says: "I believe, but am not quite sure, that

" it was on the first of March, 1837, that Mr.
" Cooke introduced himself to me. He told me
" he had applied to Dr. Faraday,* and Dr. Roget,
" for some information relative to a subject on
" which he was engaged, and they had referred
" him to me as having the means of answering
" his inquiries.   *   *   Relying upon my
" former experience, I at once told Mr. Cooke that
" *it would not, and could not, act as a telegraph,*
" *because sufficient attractive power could not be*
" *imparted to an electro-magnet interposed in a*
" *long circuit;* and to convince him of the truth of
" this assertion I invited him to King's College to
" see the repetition of the experiments on which
" my conclusion was founded.   He came, and
" after *seeing a variety of voltaic magnets which,*
" *even with powerful batteries, exhibited but slight*
" *adhesive attraction, he expressed his disap-*
" *pointment.*"†

Cooke confirms this statement by saying: " It
" was my inability *to make the electro-magnet act*
" *at long distances* which first led me to Mr.
" Wheatstone."‡

Let the difficulty of making the discovery
which overcame this *impossibility* be judged by
the fact, that for so many years, such men as
these were unable to do it when it was needed;

---

* See Appendix, Note N.   † See Appendix, Note O.
‡ See Appendix, Note P.

and let that fact answer the envious suggestion that Henry's achievement involved no great amount of analytic and inventive power.

When Barlow's demonstration was published in 1824, Henry had never seen an electro-magnet, nor tried an experiment in electricity. When, however, two years later he took up the subject, and began the first set of regular scientific investigations ever attempted in the United States, he deduced from Ampère's law the principle that the voltaic currents, carried on wires around the iron core of the electro-magnet, should move in planes at right angles to the axis of that core—which they could not do even approximately if the core itself were insulated, as in Sturgeon's small magnet, having only one coil of naked wire wound spirally around it, necessarily leaving open spaces between the successive spirals, and so leading the current like a cork screw around the core. He also reasoned that, as the current *must* be led through a spiral circuit, which theoretically *should be* circular, the departure from its true course might be counteracted by winding the wire on a second spiral outside of the first, but with its spiral angle opposed, so that the resultant of the current from the two spirals would be the same as if it revolved in planes at right angles to the axis of the core.

He brought his reasoning to the test of experiment. Instead of insulating the core, he wrapped a fine copper wire with silk, and wound it on the core; each spiral closely packed against its fellows, so as to correct the spiral error as much as possible in each layer; and then he wound the wire in a second spiral over the first, but with the pitch of the screw, so to speak, in the opposite direction. And carrying out the principle he multiplied the coils to an enormous extent in the same way. The result justified and established his theory; and his magnets at once showed a capacity hundreds of times greater than any then known to science.*

But this was not all. Another step had to be taken before Barlow's demonstration could be overthrown, and the telegraph made possible. And this he took by discovering and establishing the fact, that a magnet with a long fine wire coil must be worked by a battery of "intensity," composed of a large number of cells in series, when a *distant* effect was required; and that the greatest dynamic effect, close at hand, is produced by a battery of a very few cells of large surface, combined with a coil or coils of short coarse wire around the magnet.

These discoveries and inventions solved the problem which had seemed to European scientists insoluble; and in one account of them which was published in "*Silliman's Journal*" for January, 1831, he says : " The fact that the magnetic action

---

* See Appendix, Note Q.

" of a current from a trough is at least not sensibly
" diminished by passing through a long wire, *is*
" *directly applicable to Mr. Barlow's project of*
" *forming an electro-magnetic telegraph.*"* This
reference was to Barlow's paper of 1824, in
which he had demonstrated the impracticability
of the telegraph.

Had these things been done in the Royal Institution, and read before the Royal Society, Wheatstone
would not have been found, in 1837, denying the
possibility of an electro-magnetic telegraph ; and
Faraday would have been able to answer Cooke's
question, without sending him to Wheatstone for
the information.   In those days, however, the
United States were held in no higher estimation in
Europe, than Nazareth was in former days in
Jerusalem ; and no one in England read an
American book.

But not content with having reasoned out, and
demonstrated, that *distance* was no longer the sole
impediment in the way of the magnetic telegraph,
Henry, in 1831, established the first electro-
magnetic telegraph that ever existed.   In the
Albany academy he strung a mile of line wire,
and with an "intensity battery" at one end, and his
spool of long fine wire at the other, he operated the
armature of the first *sounding* telegraph of any
kind.   When the armature was attracted by

* See Appendix, Note R.

the magnet, it struck a small bell or sounder, which spoke its signals ; and that apparatus there was maintained to illustrate the telegraph to the students.

When applied to practical use, some code of signals must be arranged for translating the successive taps of the armature ; but that was well known in the telegraphic art for ages, needing only good judgment in arranging it, so that the letters which occur most frequently, shall be represented by the smallest number of motions ; just as Gauss and Weber arranged their needle-telegraph code in 1833, when the movements of their needle to and fro, in a number of simple combinations, indicated the alphabet.*

These "spools" of Henry have been the means by which most of the great discoveries in electro-magnetism have since been made. Faraday and Henry used them in their famous researches already referred to, in which they discovered magneto-electricity. Sturgeon, in writing of them, says : " Professor Henry has been enabled to produce a " magnetic force which totally eclipses every other " in the whole annals of magnetism; and no parallel " is to be found since the miraculous suspension of " the celebrated oriental impostor in his iron " coffin."† Without them we could not have the

* See Appendix, Note S.
† See Appendix, Note T.

telegraph, or the still more marvelous telephone.
They are to-day essentials of modern living ; and
are as familiar to us as spools of cotton.   Judg-
ing by their results, they constitute the most important discovery which has ever been made in
electricity since Volta created the battery.

Henry also put in operation at Princeton in 1835
the very simple and obvious plan of using the "in-
"tensity spool and battery," working through long
distances, to open and close the circuit of a "quan-
"tity spool and battery," stationed where the work
was to be done; thus making the powerful magnet, at short range, the servant of the weak one at
long range.   In this state he left the problem
entirely solved, to those who could procure the
money to practically apply his discoveries to the
commercial uses of man.

That task was no easy one.   In 1831 there were
no railroads and no steamships.   Over rough
country roads the mails were carried in wagons or
coaches, and the postage on a single letter was a
shilling for short distances, and twice as much for
longer.   But little capital had accumulated in this
country; and corporations, those powerful instruments for uniting the slender means of the many
into a compact force for the development of great
industrial enterprises, were hardly known.   If the
most perfect telegraph apparatus of to-day had
been then presented to the public, no company

could have been formed to exploit it. The time had not yet come; nor could it come until railroads were built, and the exchange of material things had been rendered easy.

In Europe, where money and railroads were more abundant, the telegraph was first put into practical use. Wheatstone and Cooke, in England, in 1838, having first, however, seen and talked with Henry on the subject in 1837,* *after* they had first decided the thing to be impossible, established a practical commercial telegraph line between Paddington and West Dayton, a distance of thirteen miles; and a shorter line was in Munich.† In this country private capital could not be raised for the purpose at all; not because there was any doubt that the thing could work, but because no one supposed it would repay the investment; as it certainly would not have done in those early days. At last Congress was induced to do what private enterprise refused, and in 1844, six years after the English lines had been in practical operation, and seven years after the Bavarian, money was appropriated for the line between Baltimore and Washington. This was accomplished, after great exertions, by persons hoping for the reward which a patent for some of the contrivances connected with that particular plan promised.

---

* See Appendix, Note U.

† See Appendix, Note V.

Neither in England, where Wheatstone had a patent founded on Henry's inventions, nor here, where Morse had a similar one, could the telegraph have been introduced for years after it really was, but for the beneficient operation of the patent laws. But few men are to be found who will incur the risks, and expend the money, incident to the introduction of a new and untried industry, without the hope of that pecuniary return, which in such cases, can only be secured by the exclusive use for a "limited time" of the new thing, during which it is hoped the original losses may be repaid, and a profit earned.*

Let us now consider what would have been the position of Henry in the world, if at any time before his inventions had been so long in public use that he had lost his rights, he had taken a patent for.

First, his magnetic spools, pure and simple;

Secondly, the combination of a magnetic spool of long fine wire, with an "intensity battery," for the purpose of producing a practical magnetic effect at great distances;

Thirdly, the combination with such an apparatus of a quantity battery, operating upon a spool-magnet of coarse and short wire, at a distance from the intensity battery; whereby the great lift-

* See Appendix, Note W.

ing power of the quantity magnet might be controlled by the intensity combination;

And *finally*, the combination of the intensity
battery and spool, with a vibrating armature, so
arranged as to strike a sounder when, the circuit
is closed or opened at the sending end, for the
purpose of transmitting intelligible messages tele-
graphically.

All these he might have patented in the United
States at any time during several years after his
discoveries and inventions were made; and he
could have held them against the world. That he
was the first man to do all these things is not in
doubt anywhere. If he had taken such a patent,
as late as 1837, he would have controlled the
telegraph in this country, certainly until 1851;
and unless he had then been adequately rewarded
for his great inventions, his term would have been
extended till 1858. Imagine the good he would
have done to science had the wealth which this
would have produced been poured into his purse!

But listen to his noble words: "At the time of
"making my original experiments in electro-
"magnetism in Albany, I was urged by a friend
"to take out a patent, both for its application to
"machinery, and to the telegraph; but this I de-
"clined, on the ground that I did not then consider
"it compatible with the dignity of science to confine

" the benefits which might be derived from it to the " exclusive use of any individual."

Pure science was his beloved, and he could not make merchandise of *her.*

When that sentence was written, other eminent scientists had thought differently of this question, and had patented their discoveries ; and lest he might seem to cast a reproach upon them, and to say "I am holier than thou," his humble spirit added these words: "In this, *perhaps*, I was too fastidious."

It must have occurred to him at times, when he needed money for his experiments, and when he saw the fruits of his labor enriching the world, that he *might* have taken some share of the wealth; but he would not taint with selfishness his generous gift.   How valuable in money it was he knew full well.   Even for that fragment of it, then for six years by him given to the public, which was carried to Morse in 1837 to enable him to construct his special plan of a recording telegraph in that year, now practically obsolete, Dr. Gale, who carried it, received a share in the patent which was founded upon it, and without which it could not have existed.   For that share fifteen thousand dollars in cash were subsequently paid to him.   And its use for the telegraph was but a small part of its infinite variety of applications to the arts, and the purposes of man.

Come with me now into a telegraph office, and let us see what we find there. If the line be a short one—say thirty or forty miles—you will see but one of Henry's spools, fixed to a table, having a piece of iron called an "armature," capable of vibrating in front of its poles, and so arranged that when the "spool-magnet" attracts it, it will vibrate and strike a sounding-bar of sonorous metal, which gives out distinctly the sound of the tap. The "spool" is wound spirally in layers with several hundred feet of fine copper wire, covered with silk, in the manner specified by Henry in "*Silliman's Journal.*" At the other end of the line is a battery, composed of a number of cells in series, called by Henry for distinction an "intensity battery;" and the wire circuit is supplied with a simple device, so that it may be opened or closed by the operator's finger. When he closes it, a current of electricity flows from the "intensity battery" along the wire, and around the coil of the "intensity magnet," and the armature strikes the sounder and gives the signal. The listener hears it; and as the order of the taps progresses in accordance with a pre-arranged artificial code, to express the letters of the alphabet by combinations of successive taps—just as the old visible signals were ar-

ranged by combinations of the successive movements of the vanes, or afterwards of the needle of
the Gauss and Weber telegraph—he hears letter
after letter tapped out, and the message is understood.

Now, that apparatus has nothing about it
more than was in Henry's Albany telegraph of
1831; nor could it operate if it omitted any one of
the inventions, either singly or together, which
were then for the first time combined.   It depends
entirely upon the discoveries made by Henry before
1831; and it could not have existed in the world
earlier than those discoveries, by the use of any
means then known to man ; nor since by any
other means than those discovered by Henry.

Henry used a bell as a sounder; they now use
a metal bar and a sounding box.   Henry reversed
the battery current, whereby no spring is needed
to withdraw the armature for the purpose of
vibrating it; and that is the common practice in
English and German telegraphs.   Here they
generally merely interrupt the circuit, and the
armature is withdrawn from the magnet by a
spring; although Henry's device is also used here
largely, and is essential to the quadruplex instruments.*

If, however, the telegraph line is a long one—it
may be a thousand miles or more—then you will

* See Appendix, Note X.

see two sets of Henry's spools, and two batteries. One is the "intensity battery and spool" first described; and the coil of fine wire may be, and often is, several thousand feet long—while the battery is composed of more than a hundred cells. The distance being so great they do not attempt to send force enough through the intensity circuit to operate a sounder, but only to open and close the local circuit of Henry's quantity battery and spool. That circuit consists of a battery of but one or two cells of large surface, and a spool with about a hundred feet of coarse wire wound around its core. The intensity combination opens and closes this quantity circuit, whose armature strikes the sounder, just as the intensity armature itself does on shorter lines. This obvious plan Henry described and exhibited in Princeton to his classes, long before any magnetic telegraph was ever commercially constructed, or the convenience of such an arrangement had resulted from the great length to which the lines are stretched.

Upon that apparatus there are but four names to be written. Oersted, who discovered the effect of the voltaic current upon the magnetic needle ; Arago, who discovered that the voltaic current could generate magnetism; Sturgeon, who produced the first electro-magnet; and Henry, who discovered the conditions under which an electro-magnet might be operated at a distance—who invented the devices by which it could so operate—and who applied those devices to an operative telegraph, of the same form and substance as that now in use all over the world. Beyond their discoveries and inventions nothing is essential to the present telegraph, except that which was of common knowledge when those discoveries were completed, and that ordinary mechanical skill which is far below the level either of discovery or invention.

This is the record, and so it will stand forever.

> " The moving finger writes ; and having writ,
> " Moves on : nor all your piety nor wit
> " Shall lure it back to cancel half a line,
> " Nor all your tears wash out one word of it."

Forty years had fled away since as teacher and pupil we first met, and they seemed like a dream that is past, when again we met to part forever in this world. In the chamber where the angel of death hovered over him, just ready to call him away, he talked thankfully of the past, and hopefully of that eternity on whose verge he stood. The vigor of youth and of manhood had been all spent in the service of humanity, and his strength was gone. The pallor of disease had dispelled the delicate hues of health, and time had traced its furrows on his brow. But the unclouded intellect still held its sway, enthroned in that magnificent head on which the snows of many winters had drifted ; and the gentle loving spirit still, as of old, illumined his beautiful face, but with a clearer, warmer light, reflecting the heaven upon which he gazed. For himself he had but one regret—that he had not been spared to complete his last great labor, by which he hoped to confer still one more benefit upon humanity, by discovering some means affording greater security for mariners on the treacherous coast, when fogs draw down their impenetrable veils over the lights, and the syren's voice fails to pierce the fickle air.

The faithful servant—faithful unto death—only mourned that he could not have done more. With the humble spirit of the true Christian, after having in the estimation of his fellow-men done so much, he, knowing better than others how much was yet to be done, exclaimed "I am an unprofitable servant. "

Such a life and such a death exalt and glorify humanity; illustrating and indelibly impressing upon our hearts the sublime truth, that man is made in the image of God.

" Along the smooth and slender wires, the sleepless heralds run,
" Fast as the clear and living rays go streaming from the sun ;
" No peals or flashes, heard or seen, their wondrous flight betray,
" And yet their words are quickly felt, in cities far away.

" Nor summer's heat, nor winter's cold, can check their rapid course;
" Unmoved they meet the fierce wind's blast, the rough waves
    sweeping force.

" In the long night of rain and wrath, as in the blaze of day,
" They rush with news of weal or woe to thousands far away.

" But faster still than tidings borne on that electric cord,
" Rise the pure thoughts of him who loves the Christian's life and
        Lord,
" Of him who taught, in smiles and tears, with fervent lips to pray,
" Maintains high converse, here on earth, with bright worlds far
        away.

" Aye, though no outward wish is breathed, nor outward answer
        given,
" The sighing of that humble heart, is known and felt in heaven;
" Those long frail wires may bend or break, those viewless heralds
        stay;
" But faith's last word shall reach the throne of God, though far
        away.*

* See Appendix, Note Y.

APPENDIX.

Note A, Page 11.

Life of Faraday, by Bence Jones. London: Vol. II., p 285.

Note B, Page 12.

Faraday's Experimental Researches. Vol. I., p. 1, and plate.

Note C, Page 12.

A full account of this is in Bence Jones' Life of Faraday, Vol. II., p. 1 to 6. Faraday tried it in 1824, '25, and '28, and failed each time; although, since magnetism had been developed from electricity, the converse of the problem seemed very feasible. On the 21st of September, 1831, he tried the experiment with an iron ring electro-magnet, constructed according to Henry's invention, using an intensity battery of ten pairs. One-half of the ring was wound with 72 feet of insulated wire; and the other half with about 60 feet, in the circuit of which a galvanometer was placed. When the battery was closed upon the first circuit, the iron ring became magnetic, and a current of electricity was set up, *by induction*, in the second circuit, *and the galvanometer moved.* This experiment says his biographer, is " the discovery by which he will be forever known."

Henry's account of his own discovery exhibits the very same apparatus. He used his electromagnet, capable of lifting 600 or 700 lbs., and united its poles by an iron bar or "keeper," firmly fixed, so as to form a complete circuit—the same as the iron ring in Faraday's experiment. Around this " keeper " he wound about 30 feet of insulated wire, in many layers, occupying about one inch in the length of the keeper, and placed a galvanometer in the circuit of the coil. When the battery circuit was closed and broken on the coil of the magnet, the galvanometer moved, and the great discovery was made (*Silliman's Journal*, July, 1832).

Faraday had been working over it for seven years. Henry never touched the question till 1827.

NOTE D, PAGE 21.

In another arrangement of this same invention, a heavy wire is laid between the tracks, and large inductive coils, near the floor of the car, are affected by the current in the line wire. Henry exhibited the principle of this apparatus in Princeton, when in the cellar of the Philosophical Hall, *induced* currents were set up in a wire leading around the apartment, *induced* by passing a battery current through a similar wire in the upper story, thirty feet above, and with two floors between.

NOTE E, PAGE 21.

This engine is now at Princeton in the laboratory.

NOTE F, PAGE 25.

Since the accession of President McCosh, the donations to Princeton College have been munificent. John 'C. Green, Esq., of New York, in his lifetime, and the trustees of his estate since his death, have given princely gifts, amounting to millions. The Green School of Science, with its splendid buildings and complete apparatus, supported by an endowment for its professors; the beautiful library, with an ample fund; Dickenson Hall, containing elegant lecture and class-rooms; Witherspoon Hall, one of the noblest structures in the State, and Edwards' Hall—both dormitories for students; and large sums for general purposes, are the permanent monuments of this generosity.

Henry G. Marquand, Esq., of New York, has erected a beautiful chapel, bearing his name, unsurpassed by any college chapel in this country; and in addition, has endowed a professorship of art.

Messrs. Robert L. Stewart and Alexander Stewart, brothers, gave to the College during their lifetime the splendid Potter mansion, as a residence for the President.

Mrs. Robert L. Stewart, of New York, has founded the School of Philosophy, and lately has given $150,000 to establish the chairs in this school —thus emulating the example, and equalling the generosity of her deceased husband and brother-in-law to the Theological Seminary at Princeton.

The late Mr. Hamilton Murray founded Murray Hall; and John I. Blair, Esq., of New Jersey, has endowed a professor's chair.

Messrs. Robert Bonner, Robert L. Stewart, and others, presented the magnificent 24-inch aperture telescope, which now, in the hands of the renowned astronomer, Professor Young, is doing good work, and is familiarizing students with the mysteries of the universe.

Physical culture has not been neglected by the Pactolian stream; and Messrs. Robert Bonner and Henry G. Marquand, jointly, have erected one of the most complete gymnasiums in the country.

Thus our *Alma Mater* is strengthened and adorned. Let her children, however, not forget that her great present need is scholarships, which are essential to her full development; and that even $100,000 would render an immense service in this most important department.

NOTE G, PAGE 34.

" Description of an Electric Telegraph, and some other Electrical Apparatus," by Frances Ronalds. 8 vo. London: 1823; also see *Nature*, London, November 23, 1871. Vol. 5, p. 59.

NOTE H, PAGE 34.

" The Electric Magnetic Telegraph," by Lawrence Turnbull. 8 vo., 2d Ed. Philadelphia: 1853, pp. 21, 22.

Harrison Gray Dyar, an American, set up an electric telegraph in 1827, '28, at the race course on Long Island. Static electricity was used to make a record on a strip of moving litmus paper, and the alphabet was spelled out by the intervals between the sparks passing. *Ib.*, 1st Ed., 1852, p. 6. 2d Ed., p. 22.

NOTE I, PAGE 35.

"Schweigger's Journal für Chemie und Physik." 1811. Vol. II., pp. 217, *et seq.*

NOTE K, PAGE 35.

This was Bains' English patent. The same system has been several times tried here, but with no great success. The Little-Edison " automatic " or "fast line " to Washington some years ago, worked on this plan ; and more recently the Rapid, and the Postal Telegraphs. In all cases, however, an electro-magnetic attachment is needed to give the alarm, and correct the errors.

NOTE L, PAGE 36.

Journal of the Society of Arts. July 29, 1859. Vol. VII., pp. 606–7.

NOTE M, PAGE 36.

Göttingische Gelehrte. Aug. 9, 1834. Part II., No. 128, pp. 1272–3.

NOTE N, PAGE 39.

Faraday had used Henry's quantity magnets in his experiments ; but he does not seem to have considered the effect of the combination of an intensity battery with an intensity magnet, for distant effects ; having probably overlooked Henry's demonstration of that result in Silliman's Journal of 1831.   Hence his reference to Wheatstone.

The effect of removing the magnet from the battery is stated by Daniells as late as 1843, as an elementary truth.   He says : "Electro-magnets of the greatest power, even when the most energetic batteries are employed, utterly cease to act when they are connected by considerable lengths of wire with the battery." (Introduction to the Study of Chemical Philosophy ; by Prof. John Frederick Daniell, 2d Ed., 1843, chap. XVI., sec. 859, p. 576.)

Note O, Page 39.

The Electric Telegraph : Was it invented by Professor Wheatstone ? By W. F. Cooke.   Part II.   1856.

Note P, Page 39.

Ib.   Part I., p. 19S.

Note Q, Page 41.

Professor Henry made a magnet for Yale College—still there—which lifted 2,300 pounds (Silliman's Journal, April, 1831; Vol. XX., p. 201).   His great Princeton Magnet, now in the Scientific Department, lifted above 3,000 pounds with a very small battery.

Note R, Page 42.

Vol. XIX., pp. 403–4.

NOTE S, PAGE 43.

Ingenious men for ages have amused themselves in arranging a bi-signal alphabet to obtain the most simple system. Bacon, in his great work, " On the Dignity and Advancement of Learning," Vol. VI., ch. 1, 1605, gives an alphabet of two signs, and says of it : " This contrivance shows a method of expressing and signifying one's mind *to any distance,* by objects that are either visible or audible, provided the objects are capable of two differences, as bells, speaking trumpets, fire-works and cannon," &c.

In Rees' Cyclopædia (1802–19) are given various illustrations of bi-signal and tri-signal alphabets.

Gauss and Weber's alphabets in 1833, and Steinheil's in 1836, are very nearly as simple as possible. They use, at most, only four movements, and the most frequent letters are represented by only one. Two and a-half movements to a letter are needed in the best arranged bi-signal alphabet, and these old ones probably would not require more.

NOTE T, PAGE 43.

Philosophical Magazine. March, 1832. Vol. XI., p. 199.

NOTE U, PAGE 45.

Smithsonian Report. 1857. Pp. 111–12.

NOTE V, PAGE 45.

Steinheil had his telegraph working at Munich, in July, 1837, over twelve miles of line, with eight stations. It was both a sounding and a printing telegraph, and used the earth as the return circuit. Two bells of different tones gave all the combinations needed for the alphabet. (Sturgeon's Annals, April, 1839, Vol. III., p. 520 ; Comptes Rendus, Vol. VII., pp. 590–93 ; see also Julius Dup's Anwendung des Elektro-magnetismus, Berlin, 1863, 2d Ed., 1873, sect. 5, pp. 339–347).

Steinheil's telegraph was in fact a galvanometer, in which the needle was made to swing and strike a bell, and to mark a paper by an inking apparatus —much like the cable recorder of to-day. It was necessarily weak, and quite inferior to Henry's, of 1831, in which any amount of power can be got, and a blow of any strength be delivered.

Note W, Page 46.

In the United States, patents are granted only to inventors ; but in England they are granted to those who "*introduce*" the inventions into the Kingdom, whether they are inventors or not.   The English system, although not founded on so high motives as the United States, is yet productive of more public good, because it stimulates enterprising men to seek for valuable improvements everywhere in the world and introduce them into Great Britain, where, perhaps, their authors would never come, and where, without a patent, no one would be interested in pushing them into use.   For a long time, in this country, the telegraph patent did not repay its owners for introducing it; and no one would have attempted it, or persevered in it, unless in the hope of future reward, which could not have been got unless under the protection of a patent.

When the Morse patent ran out, he had not been adequately rewarded for the expense and labor incurred in introducing his special arrangement into public use, and applied for an extension to the Hon. Charles Mason, then Commissioner of Patents. Mr. Mason consulted Professor Henry on the subject, and he advised the Commissioner to grant the extension, although he had been bitterly assailed by Morse and his friends in consequence of the testimony he had been compelled to give in the lawsuit between the owners of the patent and infringers. Professor Henry, although he thought it derogatory to the dignity of science for a *scientist* to seek for any other reward for his discoveries than the consciousness of having done good to his fellow-men, and the reputation due to his discoveries, yet fully appreciated the wisdom of the constitution and laws on the subject of patents, and thought that without such a system the discoveries of scientists, who devoted themselves to research, would never be made useful for man in those practical forms in which inventors embody them, and *introduce* them to the world, often with great labor and sacrifice.

The subjoined letter of Mr. Mason exhibits the magnanimous character of Henry. In devotion to principle, he recommended an extension of the Morse patent, knowing that to reward amply those who had introduced the invention to the world, would stimulate others to do likewise, although in doing it he specially benefited those who had "despitefully used him."

" UNITED STATES PATENT OFFICE, }
        MARCH 31, 1856.   }

"SIR,—Agreeably to your request, I now make the following statement : Some two years since, when an application was made for an extension of Professor Morse's patent, I was for some time in doubt as to the propriety of making that extension. Under these circumstances, I consulted with several persons, and among others with yourself, with a view, particularly, to ascertain the amount of invention fairly due to Professor Morse. The result of my inquiries was such as to induce me to grant the extension. I will further say that this was in accordance with your express recommendation, and that I was probably more influenced by this recommendation, and the information I obtained from you, than by any other circumstance, in coming to that conclusion.

        "I am, sir,

                "Yours very respectfully,

                        "CHARLES MASON."

"PROF. J. HENRY."

NOTE X, PAGE 50.

In reversing the galvanic current on the polarized relay used by Henry in Albany, and now largely used, the armature is moved *both ways* by magnetism—that is, it strikes the sounder, and is withdrawn from it, by magnetism. When the current is not reversed, but is broken at each signal, magnetism only operates one way, and a spring is used to withdraw the armature from the magnet, where it remains until the next magnetic impulse arrives.

NOTE Y, PAGE 54.

These beautiful and now appropriate verses appeared in a country newspaper in New Jersey in 1848. The figure was so striking, and the versification so good, that I remembered them; although I regret that the name of their author has escaped me. He was a clergyman, and had a parish in Pennsylvania.

www.ingramcontent.com/pod-product-compliance
Lightning Source LLC
Chambersburg PA
CBHW021517090426
42739CB00007B/654